Patrícia Barroso

Geostatistics in the evaluation of hydrogeological parameters of wells

AF534719

Patrícia Barroso

Geostatistics in the evaluation of hydrogeological parameters of wells

In the municipality of São Gonçalo do Amarante, Ceará

ScienciaScripts

Imprint

Any brand names and product names mentioned in this book are subject to trademark, brand or patent protection and are trademarks or registered trademarks of their respective holders. The use of brand names, product names, common names, trade names, product descriptions etc. even without a particular marking in this work is in no way to be construed to mean that such names may be regarded as unrestricted in respect of trademark and brand protection legislation and could thus be used by anyone.

Cover image: www.ingimage.com

This book is a translation from the original published under ISBN 978-613-9-68319-2.

Publisher:
Sciencia Scripts
is a trademark of
Dodo Books Indian Ocean Ltd. and OmniScriptum S.R.L publishing group

120 High Road, East Finchley, London, N2 9ED, United Kingdom
Str. Armeneasca 28/1, office 1, Chisinau MD-2012, Republic of Moldova, Europe
Printed at: see last page
ISBN: 978-620-8-21216-2

Copyright © Patrícia Barroso
Copyright © 2024 Dodo Books Indian Ocean Ltd. and OmniScriptum S.R.L publishing group

SUMMARY

ACKNOWLEDGMENTS

I thank God for giving me the health and intelligence to overcome all the difficulties and get this far.

The State University of Ceará (UECE) for giving me the opportunity to do this course.

To my mother, *Dona Conceiçao, a* warrior woman, to my father *Mauricio* (*in memori- an*), a hard-working and honest man, and to my siblings for their encouragement and unconditional support.

To my advisor, Prof. Dr. Cleyber Nascimento de Medeiros for his support, patience and excellence in guiding this work.

To my dear and eternal Professor Raimunda Olimpia de Aguiar Gomes (*in memo- rian*), my teacher who always taught me and encouraged me and whom I deeply admire and respect.

To my work colleagues who supported and motivated me.

To all my friends who encouraged me.

And to everyone who directly or indirectly played a part in my education. Thank you very much.

SUMMARY

Groundwater exploitation has been growing significantly due to the current water shortage in Ceará, caused by the fifth consecutive year of drought (2012 to 2016). It is therefore necessary to know the potential of aquifers so that guidelines for managing groundwater resources can be established based on technical criteria. In this context, this research aims to carry out a survey of existing wells in the municipality, evaluating hydrogeological parameters relating to depth, static level and flow, using Geostatistics. The methodology of this research involves the use of geostatistical procedures based on kriging techniques in the Geoprocessing environment, with the following stages being carried out: database survey; generation of thematic maps and descriptive statistical analysis; interpolation of the hydrogeological parameters of the wells and generation of kriging maps (depth, static level and flow). The *Geostatistical Analyst* extension of the Arcgis 10.2.2 program was used to generate the kriging maps. This extension offers a set of interactive tools to visually investigate the data. Ordinary kriging estimates were generated by comparing three theoretical fitting models to the experimental semivariogram: Spherical, Exponential and Gaussian. With regard to the type of use of the wells under study, it was observed that the water from the majority of wells is used for human consumption. With regard to the situation of these wells, it was noted that 49% of them are out of use. In terms of hydrogeological dominance, Sâo Gonçalo has four types of aquifer, with the crystalline system predominating. The crystalline system had the greatest average depth of 56.6 meters. The highest average static level was 9.5 meters for the Alluvial aquifer and the highest flow was identified in the barrier system, with an average of 0.5 m^3/h.

Keywords: Groundwater. Geostatistics, Kriging. Maps. Wells.

1 INTRODUCTION

The drilling of wells in Brazil's semi-arid region increased sharply with the conception of the National Sanitation Plan (PLANASA), instituted in 1969 by the Federal Government (RIBEIRO, 2013). The aim was to alleviate the deficit in the supply of water for human consumption, especially for the rural population.

As far as Ceará is concerned, it is worth noting that over the years several programs have been implemented by the federal and state governments with the aim of intensifying well-drilling efforts in order to provide the population with greater access to water.

In this context, it is worth mentioning the State Plan for Coping with Drought, drawn up by the state government in 2015, which includes a set of emergency and structural actions aimed at mitigating the adverse consequences of the drought phenomenon. Among the various actions listed in the plan is the drilling of wells.

An important aspect that should be emphasized in the aforementioned plan is its intention to strengthen the state's capacity to plan public policies in an integrated manner, embodying the idea of developing an intervention model that seeks to incorporate actions associated with living with drought in the most diverse areas, addressing actions in the areas of water sustainability, food security, economic sustainability, social benefits, knowledge and innovation (CEARA, 2015).

In relation to the drought phenomenon, it should be noted that Ceará is going through its fifth consecutive year (2012, 2013, 2014, 2015 and 2016) of drought (Figure 1), which has compromised the storage of water in surface water bodies (IPECE, 2016).

It can be seen that rainfall varies greatly over time in the state. For example, rainfall totals between 1990 and 2016 ranged from 377 mm to 1,226 mm, with 17 years recording rainfall below the historical average of 800 mm.

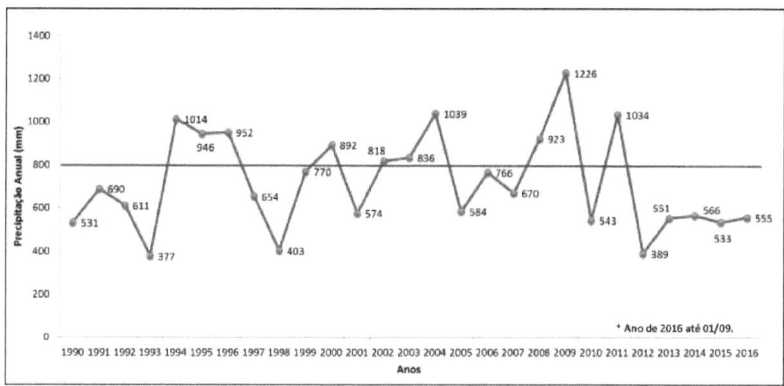

Figure 1: Average rainfall (mm) - Cearà: 1990 - 2016. Source: FUNCEME. Elaboration: IPECE, 2016. Note: The red line represents the historical average rainfall.

According to monitoring by the Companhia de Gestâo dos Recursos Hidricos (COGERH), the state's reservoirs are currently at less than 10% of their capacity (Figure 2), thus implying the need to strengthen emergency and structuring actions to guarantee the population's water security and economic sustainability.

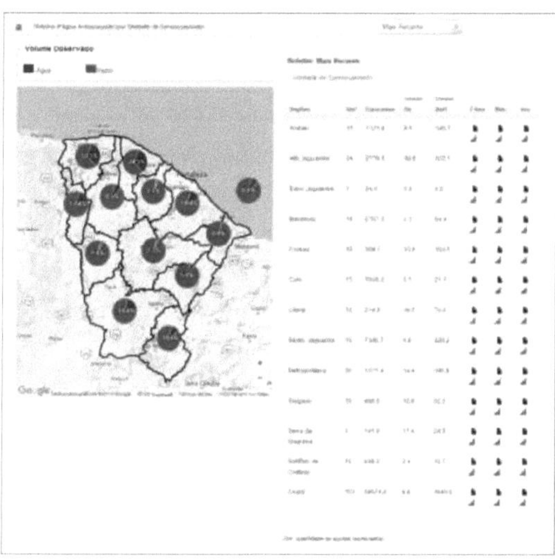

Figure 2: Volume of water stored in the state of Ceará according to river basins
Source: COGERH. Note: Position as at 30/09/2016. Note: Available at: www.hidro.ce.gov.br.

It is therefore necessary to know the potential of aquifers so that guidelines for managing underground water resources can be established based on technical criteria, capable of guaranteeing the sustainable exploitation of this resource, as well as helping to build a model for managing and granting underground water (GONÇALVES, 2009).

In this context, this research aims to carry out a survey of existing wells in the municipality of Sao Gonçalo do Amarante, evaluating hydrogeological parameters relating to depth, static level and flow using Geostatistics.

The use of statistical models has served as a tool capable of explaining the behavior of spatial phenomena that occur in nature (RAMOS et al., 2009), including the occurrence of groundwater and the potential flow rate exploitable by tube wells, using spatial interpolation techniques (RIBEIRO et al., 2010).

The municipality of Sao Gonçalo do Amarante was selected as the study area because it has a high demand for water for the development of industrial activities, as it is home to the Pecém Industrial and Port Complex (CIPP), contributing to several companies seeking to set up in the municipality, such as Companhia Siderùrgica do Pecém (CSP).

It should be noted that one of the measures of the Water Security Plan was the drilling of wells in Sao Gonçalo do Amarante to supplement the water supply of the population living on the west side of the metropolitan region of Fortaleza (RMF) and to meet the needs of the industrial sector, thus reducing the withdrawal of water from the Gaviao Dam, which receives water from the Castanhao Dam via the Eixao das Aguas.

The hypothesis is that the use of wells is a possible alternative to increase the supply of water for industrial activity and human consumption, and it is important to study the potential of underground water resources in this municipality.

2 GENERAL OBJECTIVE

To evaluate the hydrogeological parameters relating to depth, static level and flow of wells located in the municipality of Sao Gonçalo do Amarante, using geostatistical techniques.

2.1 Specific objectives

- Characterize the type of use of the existing wells in the study area;
- Analyze the distribution of wells in the municipality of Sao Gonçalo do Amarante in existing aquifer systems;
- Drawing up maps of the depth, static level and flow of the municipality's wells using geostatistics, generating subsidies for the planning and territorial management of underground water resources.

3 LOCATION OF THE STUDY AREA

Sao Gonçalo do Amarante is a municipality belonging to the Fortaleza Metropolitan Region (RMF), bordered to the north by the municipalities of Paraipaba and Paracuru; to the east by the Atlantic Ocean and the municipality of Caucaia; to the south by Caucaia and the municipality of Pentecoste and to the west by the municipalities of Sao Luis do Curu and Trairi (Figure 3). It covers an area of 834.45 km^2 , with a population of 43,890 inhabitants (IBGE, 2010), 65% of whom live in urban areas.

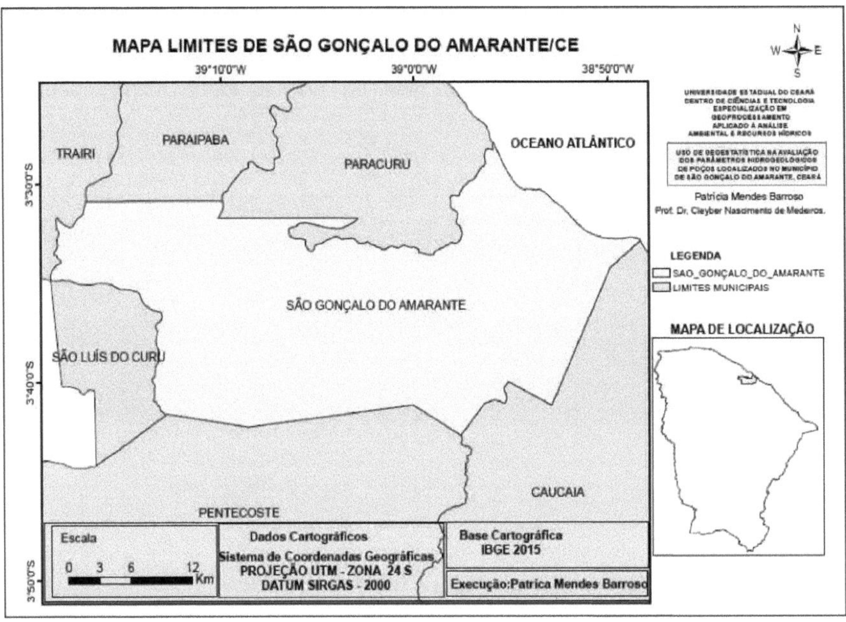

Figure 3 - Boundary Map of the Municipality of Sâo Gonçalo do Amarante (Source: Adapted from IBGE, 2015).

The climate in Sâo Gonçalo do Amarante is mild semi-arid hot tropical. Average annual temperatures range from 26° to 28°C. The rainy season is from January to May, with average annual rainfall of around 1,026 mm (IPECE, 2015). It should be noted that RMF is bordered by the Atlantic Ocean, which is influenced by the sea and consequently has milder temperatures than other regions in the interior of the state.

Figure 4 shows the isopleths for the average annual rainfall for the municipalities that were part of the RMF between 2001 and 2009, using data from the rainfall stations of the Ceará Foundation for Meteorology and Hydric Resources (FUNCEME). This map was generated using the geostatistical interpolation method of kriging.

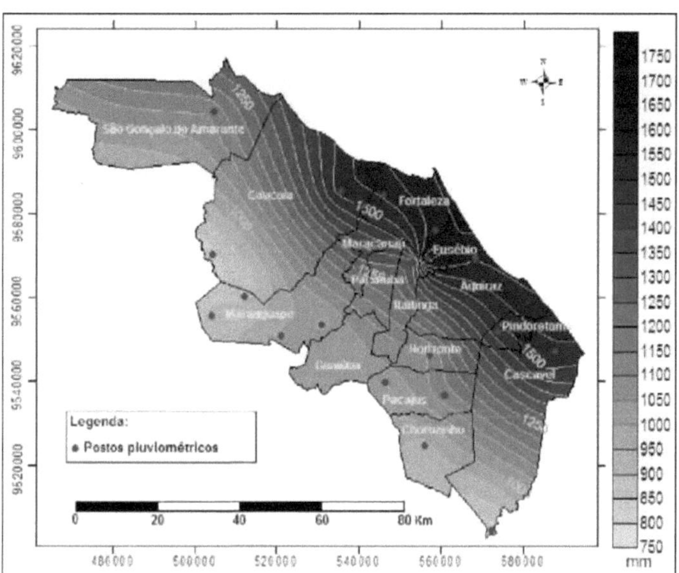

Figure 4: Isoietas of the climatological average rainfall, 2001 to 2009. Source: Magalhâes and Zanella (2011).

Rainfall is heterogeneous across the municipality. For example, in the southern part of the territory there is rainfall of around 1,000 mm, while in places near the coast there is more than 1,250 mm, including the area where the CIPP is located.

Magalhães and Zanella (2011) state that there is a higher average rainfall on the coast due to its proximity to the ocean, which favors the action of atmospheric systems in this sector, such as breeze fronts, lines of instability and easterly waves.

Thus, the climate interferes with environmental dynamics according to its climatic conditions and actions, having a direct influence on the regime and availability of both surface and sub-surface water resources. With regard to soils, the municipality of Sâo Gonçalo do Amarante is home to Red-Yellow Argisols, Quartzarenic Neosols, Lithophilic Neosols and Fluvial Neosols, which border the municipality's main rivers.

As for vegetation, the predominant types are the coastal zone vegetation complex and the caatinga (IPECE, 2015).

In hydrogeological terms, there are four aquifers in Sao Gonçalo do Amarante: Dunes/Paleodunes, Alluvium, Barreiras and Crystalline. The first three are associated with sedimentary soils, while the last is linked to crystalline soils, linked to the cracking of rocks.

4 GROUNDWATER

It should be noted that surface water resources are the main source of water supply in the state, where there are currently 153 reservoirs monitored by the Companhia de Gestao dos Recursos Hidricos (COGERH), with a storage capacity of around 18.6 million cubic meters.[3]

Nevertheless, the exploitation of groundwater has been growing significantly as a source of human and animal supply, due to the successive periods of drought that have occurred in Ceará.

Groundwater is water that occurs below the surface of the earth, filling the pores or intergranular voids of sedimentary rocks, or the fractures, faults and fissures of compact rocks, and which, being subject to two forces (adhesion and gravity), plays an essential role in maintaining soil moisture and the flow of rivers and lakes (ABAS, 2016).

An aquifer can be conceptualized as an underground geological formation capable of storing water and having sufficient permeability to allow it to move. They are real underground water reservoirs formed by rocks with porous and permeable characteristics that retain rainwater, which infiltrates through the ground, and transmits it, under the action of a hydrostatic pressure differential, so that it gradually supplies rivers and artesian wells (ECO, 2014).

According to Ribeiro (2010), aquifers can be classified according to their hydrological characteristics into free, confined and suspended aquifers:

- *Free or phreatic aquifer - is* a permeable extract, partially saturated with water, the base of which is an impermeable or semi-permeable layer. The top is bounded by the free surface of the water itself, also called the phreatic surface, under atmospheric pressure. It tends to have a profile more or less similar to that of the ground surface. The water table is usually close to the surface, in river valleys.

- *Confined or artesian aquifer* - This is an aquifer completely saturated with water, the upper (ceiling) and lower (floor) limits of which are impermeable extracts. The water in this aquifer is called artesian or confined and its pressure is generally higher than atmospheric pressure. This is why, when the aquifer is drilled, the water rises to a much higher level and can even gush out.

- *Suspended aquifers* - occur when infiltrating water encounters barriers in the unsaturated zone that can be temporarily stored.

Also according to Ribeiro (2010), aquifers can be classified according to the porosity of the storage rocks:

- *Granular* - shallow aquifers that occur in unconsolidated clastic sediments, such as river alluvial deposits, consisting mainly of sand and clay;
- *Fissures* - groundwater that occurs in rock fractures resulting from tectonic processes;
- *Karstic* - formed by carbonate rocks, which in the geological process produce fractures and fissures due to karst dissolution.

In Sâo Gonçalo do Amarante, there are the following aquifers: Crista- lino, Barreiras, Aluvioes and Dunas/Paleodunas. According to Freitas (2009), the ability of crystalline rocks to store water is intimately connected to the in- tensity, openness and interconnectedness of the fracture network, with very small values when compared to sedimentary rocks.

According to Cavalcante (1998), the flow produced by the wells averages 2.5 m^3 /h and the water, due to the lack of circulation and the effects of the semi-arid climate, is often classified as brackish. Nevertheless, this aquifer can be used as an alternative supply for small communities or as a strategic reserve during prolonged periods of drought.

The Barreiras aquifer is characterized by significant variation in clastic levels with different porosities, reflecting different potentials in terms of groundwater productivity. The depths of the wells that draw water from the Barreiras aquifer vary between 10 and 50 meters, with an average of 20 meters. Flow rates are generally small, fluctuating predominantly between an average of 3 m3/h and a maximum of 12 m3/h (CAVALCANTE, 1998).

Alluvial aquifers occur along the banks of the main rivers and streams. In the hinterland, these sediments are often the only alternative for groundwater abstraction (MEDEIROS, 2014).

According to Cavalcante (op. cit.), alluvium is a free aquifer with a thickness of up to 10 meters and a static sub-aerial level, usually less than 2 meters. They are made up of very fine-grained sediments, often interspersed with clay and organic levels, resulting from erosive action on sedimentary rocks.

The Dunes/Paleodunes hydrogeological system is made up of poorly consolidated and extremely homogeneous fine and medium sands in the area covered by the mobile and fixed dunes. According to Freitas (2009), this system has the best hydrogeological potential, representing a free-flowing aquifer, with a static level that is normally sub-affluent in the discharge areas, which are phreatic, with an average static level of 4 meters, reflecting the small thickness of these sedimentary bodies. The average flow varies between 5 and 10 m3/h, with maximums of 15 m3/h.

4.1 TUBE WELLS

The scarcity of water and the long periods of drought are driving the exploitation of groundwater. According to the Geological Survey of Brazil (CPRM), there are 22,136 registered wells in Ceará. Water from wells and springs has been used intensively for various purposes, such as human supply, irrigation, industry and leisure.

According to SOHIDRA (2016), a well is any perforation through which water is obtained from an aquifer (artesian wells), based on geological studies for the collection of groundwater through the perforation of large rocks which, when crystalline, concentrate reserves in their crevices. The types of wells vary according to the technology used, the environmental protection and safety methods, and the operating system.

Tubular wells, also known as artesian wells, are drilled using percussion, rotary and rotopneumatic drilling machines. It has an opening of a few centimeters (a maximum of 50 cm) and is lined with iron or plastic pipes (CPRM, 1998). Figure 5 shows the construction and lithology profile of a well located in Sâo Gonçalo do Amarante.

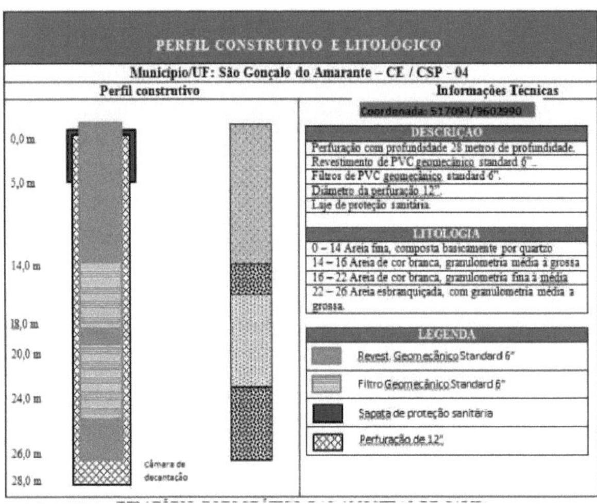

Figure 5 - Profile of a tube well. Source: SOHIDRA, 2016.

According to the CPRM (1998), when a well is drilled in a consolidated rock formation, the hole is generally kept in balance, without the need for a casing, while in a formation of sand, expansive clays, boulders and other unconsolidated formations, it must be supported by a casing or well filter in order to prevent it from collapsing or closing the well (CPRM, 1998). Also according to CPRM (1998), tube wells vary according to the hydrogeological system:

- In Crystalline Rocks:

- wells with maximum depths of around 80 meters; more often around 60 meters;
- The most common diameter is 4 to 6 inches;
- drilled with appropriate machines using percussion and compressed air;
- no need for coatings, filters or pre-filters;
- capture fissure aquifers;
- generally has low flows (average 2 to 5 m^3 /h), serving to supply homes, villages and small communities.

- In Sedimentary Rocks:

- wells of varying depths, which can reach more than 1,000 meters;
- Variable diameter, most commonly 4 to 8 inches for production coating;
- drilled with appropriate machines using percussion and rotation;
- require coatings, filters and pre-filters;
- high costs for completion materials;
- small to large flows (up to 1,000 m^3 /h);
- used to supply homes, villages, small and large communities and even populous cities.

The depth, static level and flow rate of a well vary according to the location of the aquifer in relation to the ground surface. According to Barbosa (2016), well hydraulics uses the terminology below, with its definitions:

a) Static level of the well: this is the equilibrium level of the water in the well when it is not under pumping action, nor under the influence of previous pumping, nor under the influence of pumping action taking place (or having taken place) in its immediate vicinity.

b) Depth: the distance measured from the surface of the ground to the static level of the well.

c) The pumping flow rate is the volume of water per unit of time extracted from the well by pumping equipment.

5 GEOESTATISTICS

Geostatistics assumes that the natural properties of the earth's surface are spatially continuous, requiring the application of important precepts associated with the spatial dependence of variables continuously distributed in space (CRESSIE, 1993).

According to Vasconcelos (2009), when you have a set of measurements of parameters and/or variables properly positioned, geo- statistical procedures allow you to represent the behavior of these parameters and/or variables by constructing contour maps using interpolation procedures.

Also according to Vasconcelos (op. cit), geostatistical procedures promote an investigation of the natural variations of parameters and variables with possible correlations, allowing the generation of information and cartographic representations of great value for knowledge of the potential and management of underground water resources.

According to Camargo (2001), the incorporation of geostatistical procedures into Geographic Information Systems (GIS) is important because this as- sociation improves the traditional procedures of such systems due to the quality of the estimator and, above all, the accuracy information provided in this inferential model.

This work involves the use of geostatistical procedures based on kriging techniques in a geoprocessing environment. These procedures include exploratory data analysis, semivariogram generation and modeling, model validation and interpolation using ordinary kriging.

For Lima (2006), the recognized advantages of Geostatistics over other conventional prediction techniques are:

- the study of spatial variability (the analysis of a semivariogram is the only available technique for measuring the spatial variability of a regionalized variable);
- smoothing (geostatistical estimation smoothes or regresses predicted values);
- disaggregation (or the effect of canceling out localized concentrations of observations);
- the determination of anisotropy (the behavior of variability in different directions is considered);
- accuracy (kriging provides precise values for the areas or points to be assessed);
- and uncertainty (estimates obtained through kriging include the margin of error that accompanies the estimate).

For a better understanding of geostatistics, it is necessary to understand the processes of spatial data interpolation, semivariogram, isotropy and kriging, which are presented below.

5.1 SPATIAL DATA INTERPOLATION

According to Jakob (2006), interpolation is a technique used to estimate the value of an attribute in non-sampled locations from sampled points in a study region. Spatial interpolation converts data from point observations into continuous observations, producing spatial patterns that can be compared with other continuous spatial entities.

Computational advances and improvements in mapping techniques over the last few decades have enabled us to assess the quality of mapped attributes more and more accurately, as well as to detect the errors associated with them, caused when determining the spatial representation model to be used, for example, in data interpolation (CAMARGO, 2001).

This has led to the need to implement more sophisticated ways of analyzing spatial information in Geographic Information Systems (GISs), as well as the incorporation of procedures to assess the reliability and safety of the results obtained. In the case of interpolation methods, evaluating the errors associated with the mapped attributes would be an example of this (JAKOB, 2006).

According to Jakob (2006), the most common interpolation methods used in GIS generally fall into two categories: global and local, with global methods being used more for trend surfaces, and local methods being low-order polynomials, spline functions, polyhedra, triangulation and point moving averages.

For Jakob (2006), the methods do not provide the errors associated with the estimates. Only the statistical method of kriging does this by means of a "continuous model of spatial variation". Models of quantitative attributes that are indeterminate in the field are based on averages and random average errors. The averages are stored on the maps created and these errors are defined by calculating the standard deviations. However, the standard deviation alone is usually not enough to identify the errors, but also the inclusion of autocorrelations and spatial correlations of the attributes (Jakob apud Lourenço, 1998).

5.2 THE SEMIVARIOGRAM

The semivariogram or variogram is a basic tool to support kriging techniques, which makes it possible to quantitatively represent the variation of a phenomenon regionalized in space (HUIJBREGTS, 1975). The semivariogram analyzes the degree of spatial dependence between samples within an experimental field, as well as defining the parameters needed to estimate values for non-sampled locations, using the kriging technique (SALVIANO, 1996). According to Vasconcelos, et al. (2009):

> The practice of modelling the regionalization of a property by estimating the experimental semivariogram for discrete intervals must be a prerequisite for using kriging. Therefore, plotting the contour using the kriging procedure presupposes a variogram analysis whose objective is to build a variogram model that reveals the structural features of the regionalized variable, especially when the property is anisotropic, with the variance dependent on direction and also on distance (VASCONCELOS, et al., 2009).

The semivariogram model can then be used to determine, from a limited number of observations, the best exact (unbiased) linear estimator of the property in unvisited positions.

According to Almeida (2007), the experimental semivariogram is the geo-statistical tool that allows the spatial behaviour of the regionalized variable or its residues to be assessed and demonstrates: the size of the zone of influence around a sample; anisotropy; and continuity or not at the origin (Guerra, 1998; Landim, 1998). According to Camargo (2001), the function for the semivariogram can be defined as follows:

$$\hat{\gamma}(\mathbf{h}) = \frac{1}{2N(\mathbf{h})} \sum_{i=1}^{N(\mathbf{h})} [\, z(\mathbf{u}_i) - z(\mathbf{u}_i + \mathbf{h})]^2$$

Where:

$\hat{\gamma}$ (h): is the semivariogram estimator;

h: is the distance vector (modulus and direction) between pairs of observations;

N(h): is the number of pairs, z(ui) and z(ui + h), separated by h;

z(ui) and z(ui + h): these are the values observed at locations ui and ui + h.

The semivariogram function is used to produce maps using interpolation by means of the kriging technique. Estimates are made based on values that fall below or above a certain cut-off level. Graphically, the semivariogram function is expressed as shown in Figure 6, whose characteristics are presented below:

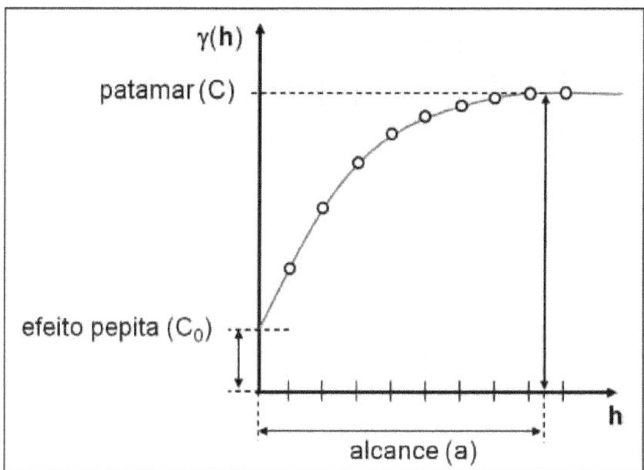

Figure 6 - Empirical (or experimental) semivariogram with characteristics very close to the ideal. Source: Camargo, 2001.

As shown in Figure 6, the experimental semivariogram has some attributes such as: Range (a), Plateau (C) and Nugget Effect (Co). These attributes are described below (author):

- Variogram amplitude (a) or range: Distance at which maximum variability is reached and which corresponds to the increase in distance between samples. In short, it corresponds to the distance within which the samples are spatially correlated;

- Plateau (c) or sill: Corresponds to the difference between the point of highest correlation or the origin of the semivariogram and the point that theoretically represents the population variance and the variability stabilizes. In short, it is the value of the semivariogram corresponding to its range (a).

- Nugget effect (co): Discontinuity at the origin of the semivariogram, corresponding to the difference between the closest samples and generated by micro-regionalization, sampling errors or measurement errors.

According to Matheron (1963), in general, the semivariogram is a function of increment with distance "h", since the further apart the samples are, the more their average values should differ. This characteristic reflects well the notion of a sample's zone of influence.

In this respect, the experimental semivariogram should be considered for a maximum of half the total sampling distance in the field, when the number of data pairs is greater than 30 (JOURNEL and HUIJBRGTS, 1991).

The notion of the zone of influence of a sample is related to the existence of regionalization. Semivariograms play a role in the spatial modeling of various phenomena, when detecting the spatial continuity or regionalization of a variable, and consequently the possible occurrence of anisotropies (Lima, 2006).

According to Lima (2006), in order to assess the behavior of a variable in spatial modeling, semivariograms are drawn up experimentally and submitted to analysis of their structural characteristics.

5.3. CONCEPT OF ANISOTROPY

Anisotropy can be seen by observing the semivariograms obtained for different directions, the most commonly used being 00, 450, 900 and 1350. Considering that the distance expressed in the semivariogram is a vector, it must be calculated for different directions. According to Lima (2006), when the semivariogram shows similar configurations for all measured directions, the phenomenon is said to be isotropic. When this is not the case, there may be anisotropy (Figure 7).

According to Camargo (2001), there are various ways of detecting anisotropy, for example, by calculating the experimental directional semivariograms in various directions, drawing them all on a single graph, and visually evaluating their similarities. Another way is through the graphical outline of an ellipse (also known as a rose diagram), calculated using the ranges obtained in different directions (DEUTSCH and JOURNEL, 1992).

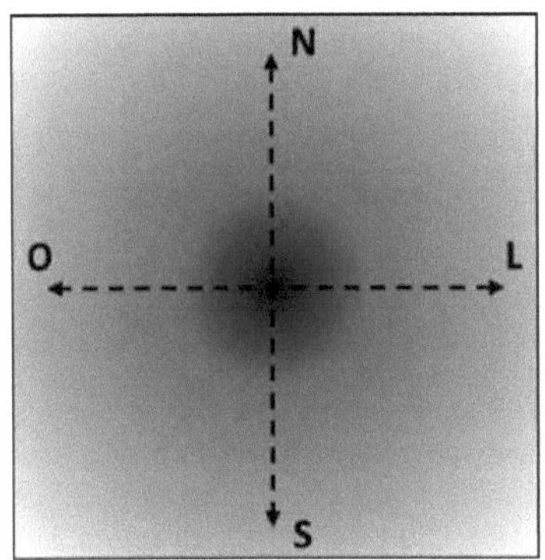

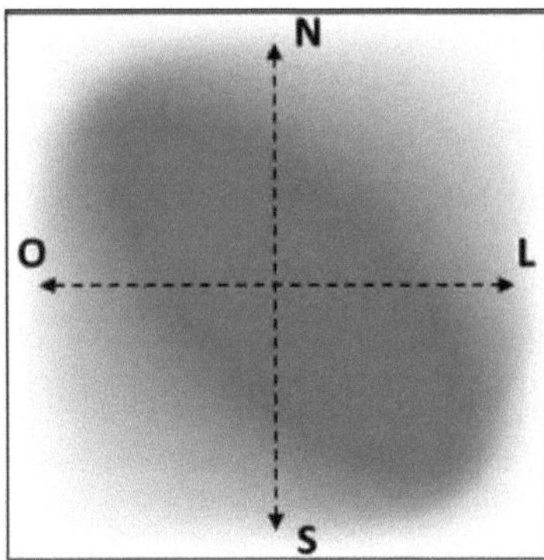

Figure 7 - Spatial variability of an isotropic and anisotropic phenomenon, respectively. Source: Camargo, 2001.

According to Camargo (2001), the most efficient and direct way of detecting anisotropy is through the semivariogram map, which is a two-dimensional graph that gives an overview of the spatial variability of the variable under study. In addition, on the semivariogram map it is possible to quickly detect the axes of anisotropy, i.e. the directions of greatest and least spatial continuity of the variable under analysis.

5.4. THE KRIGEAGE METHOD

The term krigeage is derived from the name of Daniel G. Krige, who was the first to introduce the use of moving averages to avoid the systematic overestimation of mining reserves (DELFINER and DELHOMME, 1975). What differentiates kriging from other interpolation methods is the estimation of a spatial covariance matrix that determines the weights assigned to the different samples, the treatment of data redundancy, the neighborhood to be considered in the inferential procedure and the error associated with the estimated value. In addition, kriging also provides exact estimators with unbiased and efficient properties.

Table 1 shows a comparison between deterministic and geostatistical methods, highlighting geostatistical methods in terms of dealing with redundancy, anisotropy and determining weights based on the semi-variogram, as well as the possibility of calculating estimation errors.

Table 1 - Comparison between geostatistical and deterministic methods. Source: Camargo, 2001

Geostatistical methods	Deterministic methods
> The weights are determined using a spatial correlation analysis based on the semivariogram.	> The weights are determined purely as a function of distance.
> The area of influence in interpolation is indicated by the range.	> The search radius is arbitrary.
> It models anisotropy, i.e. it detects the directions of greater and lesser spatial continuity of the phenomenon.	> Anisotropy is ignored.
> Handles redundancy ("clusters"), i.e. assigns appropriate weights to groupings of samples.	> Redundancy is ignored. In this case, overestimation or underestimation of values can occur.

Kriging methods depend on statistical models, as well as the notion of spatial autocorrelation. In classical statistics, observations are assumed to be independent, i.e. there is no correlation between observations. In geostatistics, information on spatial locations makes it possible to calculate the distances between observations and to model autocorrelation as a function of distance. The most common function used for this is the semivariogram (JAKOB, 2006).

The incorporation of geostatistical procedures into GIS, based on kriging techniques, is important because this association improves the traditional procedures of such systems due to the quality of the estimator and, above all, the accuracy information provided in this model (CAMARGO, 2001).

Once the semivariogram has been analyzed and the possibility of estimating it using geostatistical techniques has been verified, it is possible to estimate values in unsampled locations. This is an extremely important task in environmental studies, especially with regard to the spatialization and mapping of various phenomena of interest.

Kriging includes a number of estimators: Ordinary Kriging, Universal Kriging, Kriging by Indication, among others. In this work, the Ordinary Kriging estimator was used.

According to Camargo (2001), ordinary kriging is the most widely used geostatistical method in environmental studies. It consists of a linear estimation procedure for a regionalized variable that satisfies the stationarity hypothesis (constant mean of the model) and seeks to minimize the estimation error without bias, i.e. it aims for the average residual error to be close to zero. In addition, ordinary kriging has the characteristic of being the best estimator, since it minimizes the variance of the errors.

Ordinary kriging uses a random function model, based on probability, which makes it possible to assign weights to the samples used in the estimates, while also estimating the mean of the distribution of the phenomenon studied.

5.5. THE USE OF GEOESTATISTICS IN THE CHARACTERIZATION OF UNDERGROUND WATERS

According to Camargo et al. (1999) it is possible to improve the spatial distribution of environmental variables significantly when geostatistical procedures are applied.

Constanzo (2013) states that the hydrogeological model developed using geostatistical techniques in conjunction with numerical simulations of contaminant transport flows can help in decision-making in relation to environmental management in areas with contamination problems.

Another application of geostatistics is three-dimensional modeling, which has been developed with the aim of providing better representation and allows realistic and accurate numerical simulations to be carried out, which can provide subsidies for prospecting, in addition to the quantitative three-dimensional visualization itself (GOMES, 2008).

For Resende et al. (2009), the study developed using secondary data on groundwater quality parameters shows the potential of using geostatistical tools to obtain information with a measurable degree of reliability in unsampled regions.

According to Jakob (2006), interpolation methods are very good, with a high degree of data reliability, once you have the errors associated with the predicted data, but several of them require a reasonable prior knowledge of the techniques used for statistical models and their assumptions, such as prior knowledge of data normality, stationarity, trends, anisotropy, among others. The best interpolation methods in this case were determined to be ordinary kriging and simple kriging.

According to Camargo (2001), a thorough analysis using geostatistical procedures, whether for structural identification, kriging or simulation, includes a careful study of how the spatial variability of the property under analysis depends on the relative orientation of the locations of the observed data.

6 MATERIALS AND METHODS

The methodology of this research involves the use of geostatistical procedures based on kriging techniques in the Geoprocessing environment, with the following stages being carried out: Database survey; Generation of thematic maps and descriptive statistical analysis; and Interpolation of the hydrogeological parameters of the wells and generation of kriging maps (depth, static level and flow).

6.1. DATABASE

At this stage, a bibliographical review was carried out of books, articles, dissertations and theses, as well as research into the public bodies responsible for managing water resources in the state of Ceará, such as: Companhia de Pesquisa de Recursos Minerais (CPRM), using data from the Groundwater Information System (SIAGAS) and the Integrated Groundwater Monitoring Network, (RIMAS) Superintendência de Obras Hidrâulicas do estado do Cearà, (SOHIDRA), Companhia de Gestâo dos Recursos Hidricos do estado do Cearà (COGERH), as well as visits to institutions to collect data from registered wells and cartographic bases.

The well data was obtained from the well database of the Groundwater Information System - SIAGAS, provided by CPRM (SIAGAS, 2016), where 224 wells installed in the study area were identified. Subsequently, a visit was made to SOHIDRA (Superintendence of Hydraulic Works) where another 33 wells built in 2016 were registered, totaling 265 wells (Figure 8).

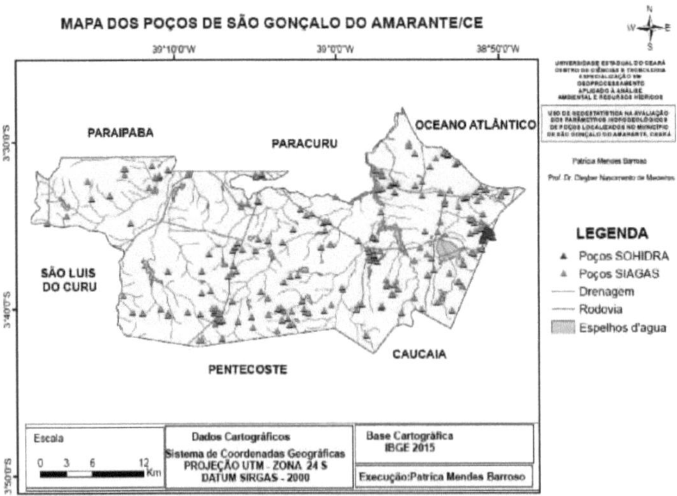

Figure 8 - Location of the wells in São Gonçalo do Amarante. Source: Prepared by the author.

The well data was catalogued in an electronic spreadsheet for better interpretation and analysis. The spreadsheet contained fields referring to UTM coordinates, municipality, location, date the well was built, depth, static level (NE), flow rate, situation and type of use.

A georeferenced database was created in a Geographic Information System (GIS) with cartographic data, thematic maps, satellite images, coordinates, shape files, among others, after which the existing parameters were tabulated and the data integrated, where mapping made it possible to identify the distribution of wells over the four aquifer systems in São Gonçalo do Amarante.

The hydrogeological parameters relating to depth, static level and flow of the wells located in the municipality were analyzed using geostatistical techniques and it was verified which aquifer system is most suitable for drilling new wells.

6.2. generation of thematic maps and descriptive statistical analysis

To create the thematic maps, we used georeferenced files of the boundaries of Ceará's municipalities, available on the IPECE website, a SPOT-5 satellite image from the Ceará Institute for Research and Economic Strategy (IPECE) to generate the hydrogeological map of Sâo Gonçalo do Amarante, and coordinates of the existing wells on the SIAGAS system website, as well as the database of wells registered by SOHIDRA.

The coordinate system used was the Universal Transverse Mercator (UTM) projection system, SIRGAS 2000 datum. The software used was Arcgis 10.2.2.

6.3. INTERPOLATION OF HYDROGEOLOGICAL PARAMETERS AND GENERATION OF KRIGEAGE MAPS (DEPTH, STATISTICAL LEVEL AND FLOW)

The *Geostatistical Analyst* extension of the Arcgis 10.2.2 program was used to generate the kriging maps. This extension offers a set of interactive tools for visually investigating the data, covering aspects relating to: descriptive statistics, histogram graphs and data normality, chloroplastic maps, making the semivariogram, assessing isotropy, selecting the kriging method and model.

Ordinary kriging estimates were generated by comparing three theoretical fitting models to the experimental semivariogram: Spherical, Exponential and Gaussian, and the best model was chosen based on the calculation of the Spatial Dependence index, as well as by analyzing the parameters relating to the Mean Square Error (MSE), which should have a value close to zero, thus indicating a good fit.

The Spatial Dependence Index (SDI) was proposed by Zimback (2001) with the aim of capturing the spatial dependence of data by quantifying the extent to which the data is spatially correlated. The SDI is calculated according to the following formula:

$$IDE = \frac{C_1}{C_1 + C_0} \times 100$$

Where:

Nugget effect (C0);

Level (C0+C1) e;

Contribution (C1).

In this context, the relationship is as follows: FDI values of less than 25% imply weak spatial correlation; FDI values between 25% and 75% correspond to moderate spatial correlation; and FDI values greater than 75% imply strong spatial correlation.

It should be noted that the use of geostatistical methods is only justified when there is strong or moderate spatial correlation in the data. Otherwise, it is more appropriate to use deterministic methods, such as the inverse of the square of the distance.

7 RESULTS

Table 1 shows descriptive statistics for the parameters relating to depth, static level and pumping flow of the existing wells in Sao Gonçalo do Amarante, based on the SIAGAS database and SOHIDRA well data. The average depth of the wells is 44.30 meters, reaching a maximum of 90 meters. The estimated average static level for the municipality of Sao Gonçalo do Amarante is 7.05 meters, with a median of 6.43 meters. The average flow rate was 4.94 m^3 /h, with a standard deviation of around 15.64 m3/h.

Table 1 - Descriptive statistical summary of the parameters studied

Statistics	Depth (meters)	Static level (meters)	Flow (m3/h)
Average	44,30	7,05	4,94
Median	52,00	6,43	0,23
Fashion	60,00	9,00	0,08
Maximum	90,00	29,00	34,00
Minimum	3,65	0,90	0,003
Standard deviation	26,16	9,19	15,64

Source: Prepared by the author.

Of the wells registered in the SIAGAS system, 51% are equipped, 13% are closed, 22% have not been identified, 9% have not been installed, 4% are abandoned and 1% are dry (Graph 1). With regard to current use, 44% of the wells are for domestic use; 37% have multiple uses (domestic/irrigation or domestic/livestock); 5% for livestock, 5% were not identified, 4% for urban supply, 3% for irrigation and only 2% are for industrial supply (Graph 2).

Graph 1 - Status of wells registered with SIAGAS.

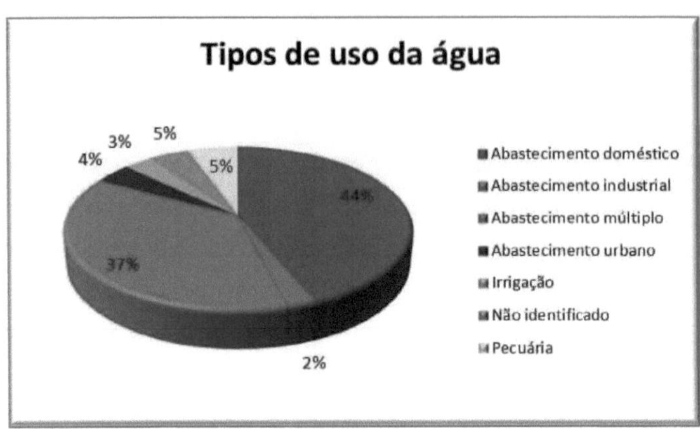

Source: Prepared by the author.

Graph 2 - Use of wells registered with SIAGAS.

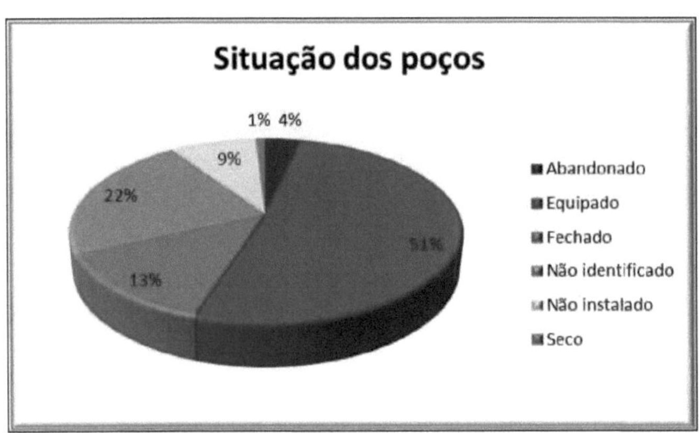

Source: Prepared by the author.

The time span of wells registered in the SIAGAS system ranged from 1940 to 2012, with no wells registered in the system in the last five years. However, according to data from the Secretariat of Water Resources (SRH), 1,150 wells were built in Ceará in 2015. In 2016, until August, the number reached 1,139 units. You can see, therefore, that the number

of existing wells is far greater than the number of wells currently registered with SIAGAS.

It should be noted that the main objective of the wells drilled by SOHIDRA is to identify alternative sources of water, and that water will be used for human consumption and industrial activity, in an attempt to avoid rationing in the Fortaleza metropolitan region.

It should also be noted that the 33 wells drilled in the Pecém region were built in the porous

Dunas and Barreiras aquifer. The average depth was 21.08 meters, ranging from 14 to 28 meters. The average static level was 3.8 meters with values between 0.9 and 9 meters and a flow rate of 15.69 m^3/h.

Map 1 shows the classification of the aquifers found in São Gonçalo do Amarante, showing that the Crystalline Aquifer is the largest, followed by the Barreiras, Dunes/Paleodunes and Alluvial aquifers. There is also a greater number of wells located in the Cristalino aquifer, above all because it has the largest territorial area.

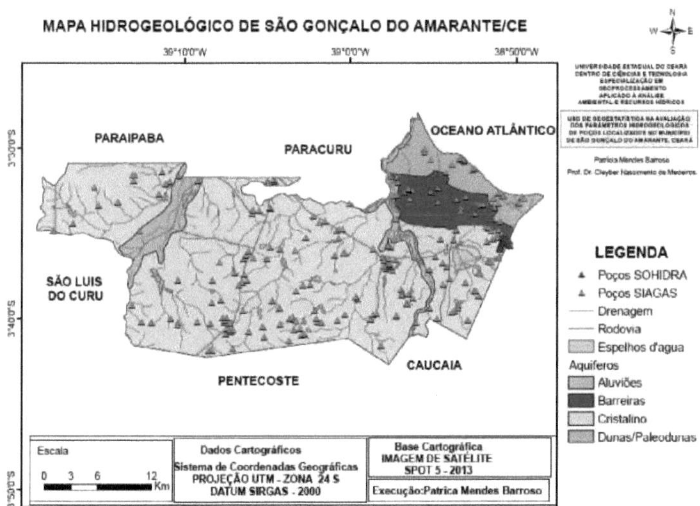

Figure 9 - Classification of the aquifers present in São Gonçalo do Amarante: Source: Adapted from the State Environmental Systems Map using Spot 5 satellite image, year 2013, scale 1:100.000.

With regard to the distribution of these wells by hydrogeological domain, it was found that there are 7 wells in the Alluvial System, 23 wells are in the Dunes/Paleodunes System, 24 in the Barreiras System and 170 in the Crystalline System (Graph 3).

Graph 3 - Percentage of wells by type of aquifer in São Gonçalo do Amarante/Ce

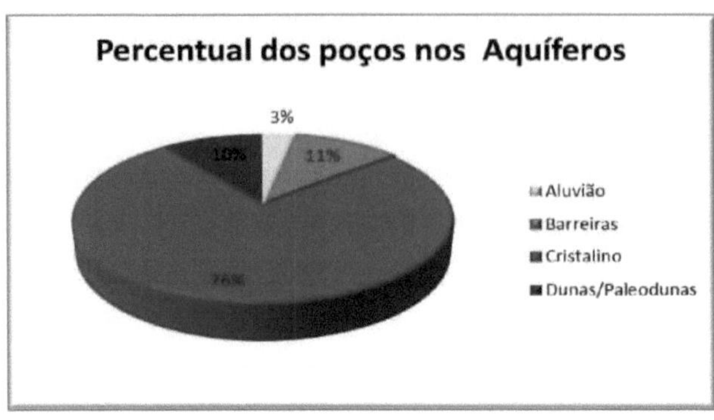

Source: Prepared by the author.

Considering the wells separately by aquifer type, there is an average static level of 5.38 meters for the Barreiras System, with values between 2.4 and 10 meters; 7.3 meters for the Dunes System, with values between 2 and 25.07 meters; 9.5 meters for the Alluvial System, with values between 3.5 and 16 meters; and 8.6 meters for the Crystalline System, with values between 2.2 and 29 meters.

The crystalline system had the highest average depth of 56.6 meters, with maximums and minimums of 90 and 3.6 meters. The Du- nas/paleodunes system had the lowest average depth, 17.9 meters, with values between 50 and 4.2 meters. The greatest flow was in the barrier system with an average of 0.5 m^3/h, with maximums and minimums of 1.4 and 0.02 m^3 /h (Table 2).

Table 2 - Average parameters studied in the aquifers

Aquifers	Depth (m)	Static Level (m)	Flow (m^3 /h)
Alluvium	46,7	9,5	0,09
Barriers	22,5	5,4	0,5
Crystal	56,6	8,6	0,3
Dunes/Paleodunes	17,9	7,3	0,2

As for the type of use in the crystalline aquifer, 44% is for multiple supply and 38% for domestic supply; in the barrier aquifer, 62% is used for domestic supply; in the Dunas/Paleodunas aquifer 74% is used for domestic supply; in relation to the alluvial aquifer there was an average of 29% for domestic, multiple and urban supply.

With regard to the status of these wells, in the alluvial aquifer 43% are equipped; in the barrier aquifer 79% are equipped; in the dunes/paleodunes 48% are equipped and in the crystalline aquifer 49% are equipped. The remainder is divided into closed, not installed and abandoned, with a large number of disused wells. Another important factor is the lack of completed data, since not all wells have complete data in the SIAGAS system.

The following maps show the results of the geostatistical kriging technique for the parameters relating to depth, static level and flow of the wells registered in the SIAGAS system in the municipality of Sâo Gonçalo do Amarante.

The first parameter studied was the depth of the wells. Figure 10 shows a wide variation in depth, with the shallowest wells (blue on the map) in the northeast of the municipality, specifically in the area of the barrier and dune/paleodune systems. In turn, the deepest wells are located in areas of the sertanejo depression.

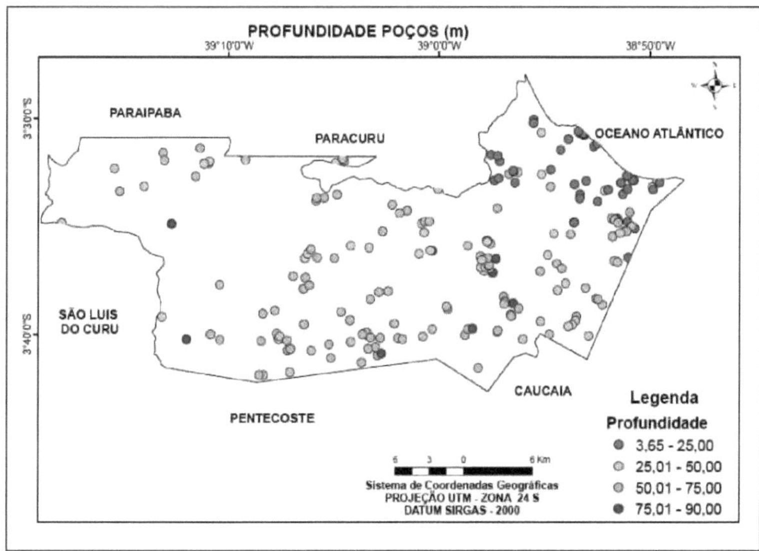

Figure 10 - Depth map of the wells in Sâo Gonçalo do Amarante/CE.

Figure 11 shows the map with the estimated depth of the wells using geostatistical interpolation. The best model corresponded to ordinary kriging using the exponential function, which had a mean square error of 13.73 and an index of spatial dependence equivalent to 84%. The parameters used in the interpolation, such as the number of observations, which in this case corresponded to 203 wells, can be found in the appendix.

In summary, the wells located in the northeast of the municipality require less depth to collect water. There is also a region located in the crystalline aquifer which has depths varying between 23 and 37 meters, while most of this system has depths of more than 50 meters.

Figure 12 shows the map with the estimation errors, indicating that the estimates with the highest errors are located in places where there are no wells.

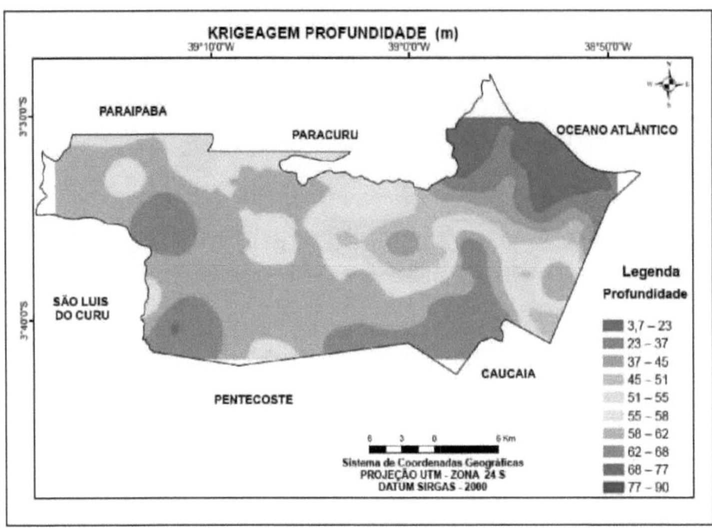

Figure 11 - Kriging map of the well depth parameter.

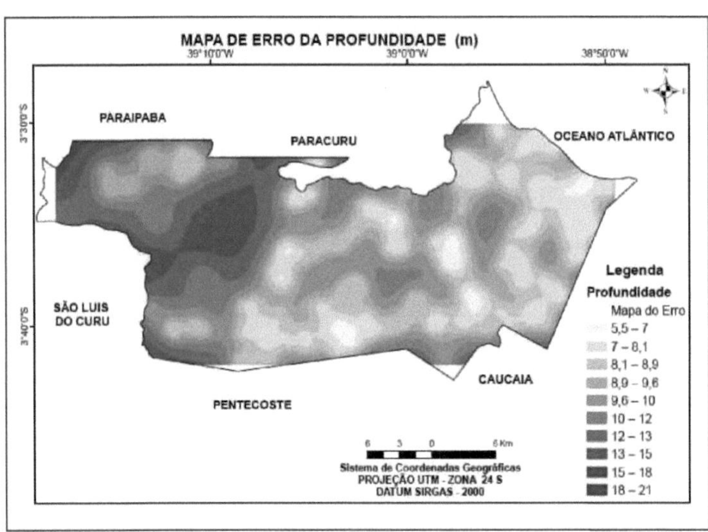

Figure 12 - Depth parameter error estimation map.

The second parameter evaluated refers to the static level of the wells (Figure 13), with values varying between 2 and 29 meters. However, there is a prevalence of wells with a static level between 2 and 5.5 meters.

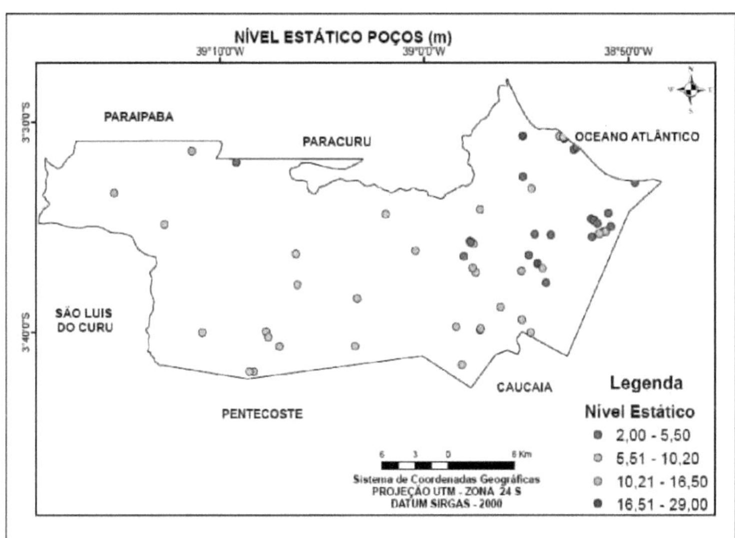

Figure 13 - Map of the static level of the wells in Sâo Gonçalo do Amarante/CE.

Figure 14 shows the map with the static level indicator of the wells using geostatistical interpolation. It should be noted that for this parameter only 61 wells had information, which hindered the estimation of the parameter for non-sampled sites. Within the context of data limitations, it was found that the model with the best estimates was ordinary kriging using the spherical model, which had a mean square error of 4.71 and an index of spatial dependence equivalent to 76%. The parameters used in the interpolation can be found in the appendix.

An analysis of Figure 14 shows that the lowest static levels of the wells are located in the region of the barrier aquifers and dunes/paleodunes. It is interesting to note the heterogeneity present in the region of the sertanejo depression, with places having values below 5 meters and others with values above 15 meters. Figure 15 shows the error estimation map.

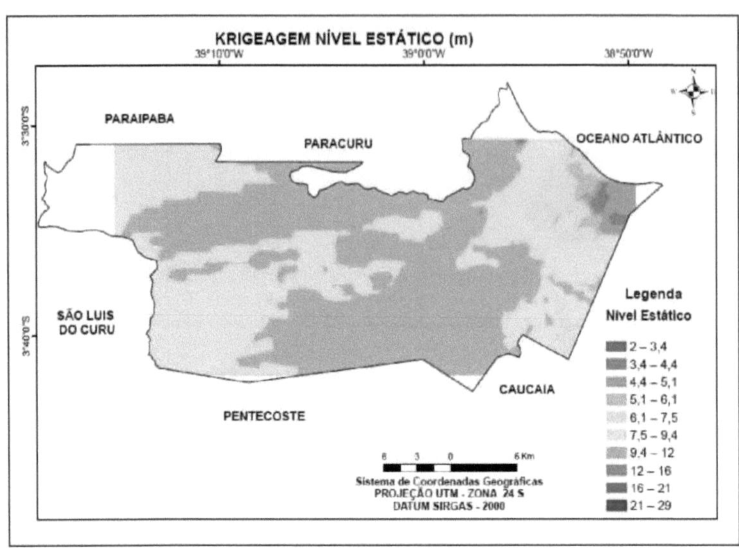

Figure 14 - Kriging map of the static level parameter of the wells.

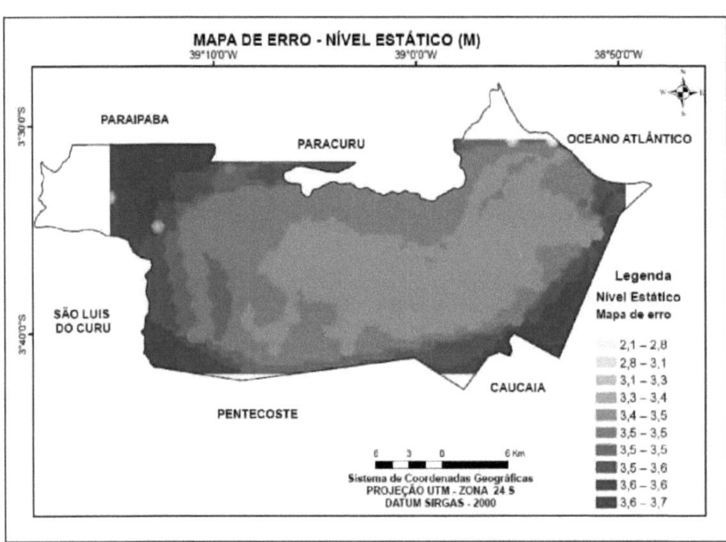

Figure 15 - Map of the error estimate for the static level parameter.

The third and final parameter studied was the flow rate of the wells, with values varying between 0.3 and 1.4 m^3 /h. It should be noted that there was a prevalence of wells with values below 0.157 m3/h (Figure 16).

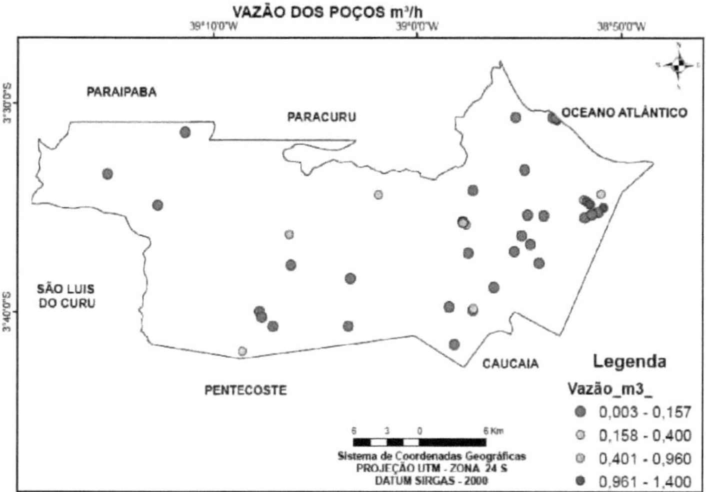

Figure 16 - Map of the static level of the wells in Sâo Gonçalo do Amarante/CE.

Figure 17 shows the map with the indicator of the flow level of the wells using geostatistical interpolation. It should be noted that for this parameter only 48 wells had information, which hindered the estimation of the parameter for non-sampled sites. Within the context of the limited data, the model with the best estimates corresponded to ordinary kriging using the Gaussian function, which had an average quadratic error of 0.35 and a spatial dependence index of 60%. The parameters used in the interpolation can be found in the appendix.

An analysis of Figure 17 shows that the lowest well flow rates are concentrated in the southern region of the municipality (blue on the map), located in the sertaneja depression (crystalline aquifer). It is interesting to note that even in the sertanejo depression, significant flows were recorded.

In turn, the highest well flow estimates are found in the barrier and dune/paleodune aquifers. Figure 18 also shows the error estimation map for the well flow indicator.

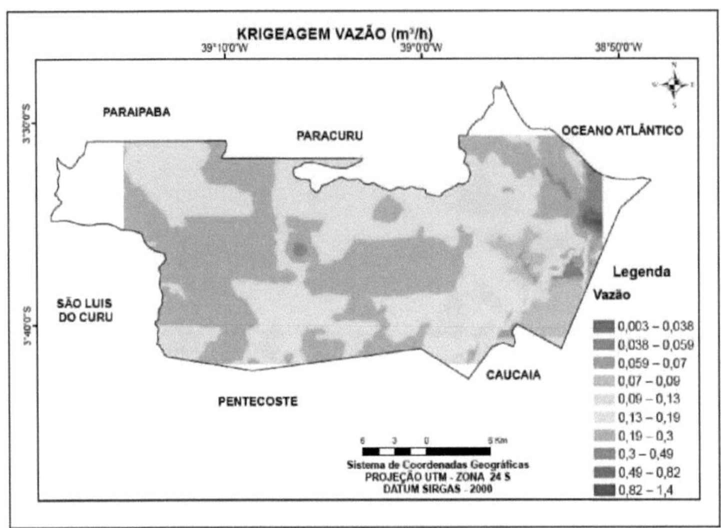

Figure 17 - Kriging map of the well flow parameter.

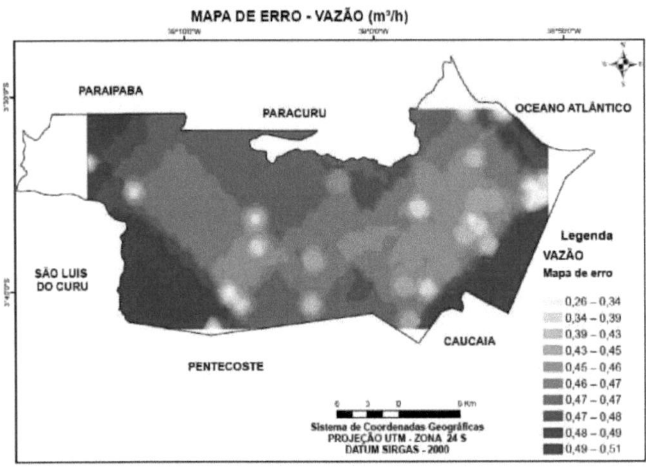

Figure 18 - Map of the error estimate for the well flow parameter.

8 FINAL CONSIDERATIONS

This study evaluated the hydrogeological parameters relating to depth, static level and flow of wells located in the municipality of Sao Gonçalo do Amarante, using geostatistical techniques. Ordinary kriging estimates were generated by comparing three theoretical fitting models to the experimental semivariogram: Spherical, Exponential and Gaussian. The maps showed in which aquifer system the greatest depths, static levels and flows were located, and maps of estimation errors were also generated, indicating that the estimates with the greatest errors are found in places where there are no wells.

The methodology used proved to be efficient because, using the Ordinary Kriging method, it was possible to see in which aquifer systems it is more likely to build new tube wells, highlighting that this spatialization technique can be an instrument capable of assisting the planning and management of underground water resources.

Ordinary kriging estimates were generated using the spherical model for static level, the exponential model for depth and the Gaussian model for the flow map. The best model was chosen by calculating the Spatial Dependence Index (SDI), as well as by analyzing the parameters relating to the Mean Square Error (MSE), which should be close to zero, thus indicating a good fit. The SDI values were around 60 to 80%, indicating strong or moderate spatial correlation in the data.

This study showed the importance of using geostatistical tools to obtain information with a measurable degree of reliability in non-sampled regions. The error was close to zero, indicating good estimates for the parameters evaluated.

Regarding the type of use of the wells in Sao Gonçalo do Amarante, it was observed that the majority of wells are used for human consumption. With regard to the situation of these wells, it was found that 49% of them are out of use.

In terms of hydrogeological dominance, there are four types of aquifers in São Gonçalo, where the crystalline aquifer predominates, which has a low hydrogeological potential, characterized by low flows and high depths. It is in this context that most of the municipality's registered tube wells are located (170 of the 224 wells). It was also observed that the alluvial system is little explored, with only 7 registered wells.

With regard to the depth variable, the crystalline system had the highest average with a value of 56.6 meters and the dunes/paleodunes system had the lowest average depth, 17.9 meters, which is more conducive to drilling new wells.

The highest average static level was 9.5 meters for the Alluvial aquifer, with minimum and maximum values of 3.5 and 16 meters. The lowest average was recorded in the barrier system, with a value of 5.4 meters.

The highest flow rate was found in the barrier system, with an average of 0.5 m^3/h, with highs and lows of 1.4 and 0.02 m^3/h. The lowest was 0.09 m^3/h in the alluvial system. It should be noted that the highest flow rate indicates the greatest potential of a well as a water producer.

An important fact observed during the course of this work is that the wells registered in the SIAGAS system are out of date, since the last registered well was drilled in 2012, but according to research, new wells have already been built over the last five years. It was also noted that some parameters did not have all the necessary information, and it is essential that SIAGAS data is updated at least annually.

Finally, in addition to drilling new wells, it is important to restore wells that are out of use. It is recommended that these wells should be included in the State and Municipal Government's recovery and installation programs to increase the municipality's water supply.

9 REFERENCES

ABAS. (2016). **Brazilian Groundwater Association.** [online]. Sao Paulo - SP. Available at: http://www.abas.org/educacao.php. Accessed on: October 20, 2016.

ALMEIDA, Ronaldo de; et al. **Geostatistical method for environmental modeling of pollutants in lake systems - Western Amazon.** Anais XIII Simpòsio Brasileiro de Sensoriamento Remoto, Florianópolis, Brazil, April 21-26, 2007, INPE, p. 2247-2253.

BARBOSA, Antenor Rodrigues Júnior. **Groundwater - Well Hydraulics**.

Elements of Applied Hydrology. Available at:

http://www.leb.esalq.usp.br/disciplinas/Fernando/leb1440/Aula%208/Hidraulica%20d and%20Pocos_Anteor%20R%20Barbosa%20Jr.pdf. Accessed on: January 17, 2017.

BOMFIM, Luiz Fernando Costa. **MAP OF HYDROGEOLOGICAL DOMAINS/SUBDOMAINS OF BRAZIL IN SIG ENVIRONMENT: CONCEPT AND METHODOLOGY.** XVI Brazilian Congress of Groundwater and XVII National Meeting of Well Drillers. 2010.

CAMARGO, E. C. G. (2001). **Geostatistics: fundamentals and applications.** Available: http://www.dpi.inpe.br/gilberto/tutoriais/gis_ambiente/5geoest.pdf. Accessed on: October 2016.

CAMARGO, E. C. G. et. Al. (2001). **The Importance of Modeling Anisotropy in the Spatial Distribution of Environmental Variables Using Geostatistical Procedures.** Anais X SBSR, Foz do Iguaçu, April 21-26, 2001, INPE, p. 395-402. Oral Technical Session.

Integration of Geostatistics and Geographic Information Systems: A Necessity. [CD-ROM]. In: V Congress and Trade Fair for Geoprocessing Users in Latin America, 7, Salvador, 1999. **Proceedings**. Bahia, gisbrasil'99. Technical-Scientific Lectures Section.

CAVALCANTE, I. N. **Fundamentos Hidrogeológicos para a Gestao Integrada de Recursos Hidricos na Regiao Metropolitana de Fortaleza - Estado do Cearà**. Doctoral Thesis - IG/USP. Sao Paulo - SP. 156 p. 1998.

CAVALCANTE, I. N.; MORAIS J. B. A. de; SOUSA V. P. de; ARAÙJO K. V. de; GOMES M. da C. R. & MATTA. M. A. da S. **THE UNDERGROUND WATERS OF SÀO GONÇALO DO AMARANTE, WEST COASTAL BELT OF THE METROPOLITAN REGION OF FORTALEZA - CEARA.** *II International Congress on the Underground Environment*. 2011.

CEARA - **State Plan for Coping with Drought**. Fortaleza. 2015. Available at: <

http://www.ipece.ce.gov.br/index.php/publicacoes>. Accessed on: 22 Nov. 2016.

Cearà. Ceará State Legislative Assembly. **Current scenario of water resources in Cearà / Council for High Studies and Strategic Affairs, Legislative Assembly of the State of Cearà;** Eudoro Walter de Santana (Coordinator). - Fortaleza: INESP, 2008. 174p. : ill. - (Water Pact Collection).

COGERH. **Dam levels.** Available at: http://www.hidro.ce.gov.br/. Accessed on: September 30, 2016.

COSTANZO, Caetano Pontes; VIDAL, Alexandre Campane, GIMENEZ, Marina Marques & Simony Yumi Yaginuma Sakamoto. **GEOESTATISTIC HYDROGEOLOGICAL MODEL FOR NUMERICAL SIMULATION OF CONTAMINANT FLOW AND TRANSPORT.** III International Congress on the Subterranean Environment, UNICAMP, Institute of Geosciences, Department of Geology and Natural Resources. 2013.

CPRM. Geological Survey of Brazil (2016). **Groundwater Information System (SIAGAS).** Available at: http://siagasweb.cprm.gov.br/layout/pesquisa_complexa.php. Accessed on October 12, 2016.

CPRM/Ministry of Mines and Energy - Geological Service of Brazil. 1998. **Basics of tube wells**. Brasilia, 21p.

CRESSIE, N. A. C. **Statistic for spatial data**. New York: J. Wiley, 1993. 900p.

DA SILVA, F. J. A., ARAÙJO, A. L. de. SOUZA, R. O. **Aguas subterrâneas no Cearà - poço instalados e Salinidade.** *Rev. Tecnol. Fortaleza, v. 28, n. 2, p. 136-159, dec. 2007.*

DEUTSCH, C.V.; Journel, A. G. GSLIB: **Geostatistical Software Library and user's guide** . New York, Oxford University Press, 1992. 339p.

ECO. **Environmental journalism.** What is an aquifer? Available at: http://www.oeco.org.br/dicionario-ambiental/28001-o-que-e-um-aquifero/. Accessed on: Oct. 2016.

FREITAS, L. C. B. **Quality of groundwater - Area in the municipality of Caucaia, Metropolitan Region of Fortaleza, Cearà**. Master's dissertation. Postgraduate Program in Geology, UFC, 110 p. 2009.

GONÇALVES, T D.; ROIG, H. L.; CAMPOS, J. E. G Geographic information system as a tool to support the allocation of groundwater resources in the Federal District. **Revista**

Brasileira de Geociências, Sao Paulo, v. 39, n. 1, p. 169-180. 2009.

GOMES, Moab Praxedes, VITAL, Helenice & Macedo, José Wilson de Paiva **Application of geostatistics in the filtering of bathymetric and altimetric data in the Potiguar Basin**. Revista de Geologia, Vol. 20, n° 2, 243-254, 2008.

GUERRA, P. A. G. Operational geostatistics. Brasilia: National Department of Mineral Production. 145 p.1988.

IBGE. Brazilian Institute of Geography and Statistics. **IBGE CITIES**. Available at: http://www.cidades.ibge.gov.br/xtras/perfil.php?lang=&codmun=231240&search=cear a|sao-goncalo-do-amarante. Accessed on: October 20, 2016.

IPECE - Instituto de Pesquisa e Estratégia Econòmica do Cearà. **indice Municipal de Alerta 2016.** Fortaleza. 2016. Available at: <www.ipece.ce.gov.br/categoria4/ima/>. Accessed on: November 23, 2016.

IPECE - Instituto de Pesquisa e Estratégia Econòmica do Cearà. **Municipal Basic Profile**. Sao Gonçalo do Amarante. 2015. Available at:

<http://www.ipece.ce.gov.br/publicacoes/perfil_basico/index_perfil_basico.htm>. Accessed on: November 23, 2016.

JAKOB, Alberto Augusto Eichman. YOUNG Andrea Ferraz. **The use of spatial data interpolation methods in sociodemographic analysis**. XV National Meeting of Population Studies, ABEP, held in Caxambu - MG - Brazil, from September 18 to 22, 2006.

JOURNEL, A.G.; HUIJBREGTS, C.J. **Mining Geostatistics**. New York: Academic Press, 1978.

LIMA, Prof. Eduardo Rodrigues Viana de. **GEOESTATISTICA APLICADA AO ESTUDO DE SOLOS** (Mini course). Department of Geosciences Federal University of Paraiba. PB, 2006.

LOURENÇO, R.W**. Comparison of interpolation methods for Geographic Information Systems**. Master's dissertation prepared for the Postgraduate Course in Geosciences - Geosciences and Environment concentration area at the Institute of Geosciences and Exact Sciences of the São Paulo State University, Rio Claro Campus. Rio Claro, SP, 1998.

MAGALHAES, G. B; ZANELLA, M. E. Climatic behavior of the Metropolitan Region of Fortaleza. **Revista Mercator**. v. 10, n° 23, p. 129-145. 2011.

MANOEL FILHO, J. **Occurrence of groundwater**. In: FEITOSA, F A. C.; MANOEL FILHO, J. (Ed.). *Hydrogeology*: concepts and applications. Fortaleza: CPRM, 2000. p. 13-33.

MATHERON, G. Principles of geostatistics. **Economic Geology**, v. 58, p.12461266. El Paso, 1963.

MEDEIROS, C. N. **Vulnerabilidade Socioambiental do municipio de Caucaia (CE): Subsidios ao ordenamento territorial.** Doctoral thesis. Postgraduate Program in Geography. UECE. Fortaleza - CE. 267 p. 2014.

MOURA P.; SABADIA, J. A. B.; CAVALCANTE I. N. **VULNERABILITY mapping OF DUNE, BARRIER AND FISSURE AQUIFERS IN THE NORTH PORTION OF THE PECÉM INDUSTRIAL AND PORT COMPLEX, STATE OF CEARA.** *Sao Paulo, UNESP, **Geosciences**, v. 35, n. 1, p.77-89,* 20178 *2016.*

MOURA, P.; CAVALCANTE, I. N.; SABADIA, J.A.B.; MORAIS Joao B. A. de. **Characterization of groundwater abstraction and use in the Pecém Industrial and Port Complex, Ceará-Brazil.** Revista de Geologia, Vol. 26, n° 1,61 - 72. 2013.

Ceará's Water Resources: **Integration, Management and Potential.** Cleyber Nascimento de Medeiros, Daniel Dantas Moreira Gomes, Emanuel Lindemberg Silva Albuquerque, Maria Lùcia Brito da Cruz (Organizers). Fortaleza: IPECE, 2011. 268 p. ISBN: 978-85-98664-20-0.

RAMOS, C. M. C. et al. Temporal Analysis of Temperature Variation Using Geostatistics. In: Simpòsio Brasileiro de Sensoriamento Remoto, 14, 2009, **Anais...** Natal, Brazil, INPE, p. 347-353. 2009.

RESENDE, Ronaldo Souza. **Atlas of groundwater quality in the state of Sergipe for irrigation purposes** / Ronaldo de Souza Resende, Marcus Aurélio Soares Cruz, Julio Roberto Araujo de Amorim. - Aracaju : Embrapa Tabuleiros Cos- teiros, 2009. 46 p. : ill. color.

RIBEIRO, D. D. M.; ROCHA, W J. S.F.; GARCIA, A. J. V. **Definition of potential areas for the occurrence of groundwater in the siriri river sub-basin - Sergipe with the aid of AHP (analytical hierarchy method).** In: CONGRESSO BRASILEIRO DE AGUAS SUBTERRÂNEAS, 16, 2010, **Anais...** Sao Paulo, Brazil - **eIssN 2179- 9784**, 2010.

RIBEIRO, P. H. B. **Spatialization of the flow produced by tube wells in different hydrogeological formations in northeastern Bahia.** 2013. 90f. Dissertation (Master's Degree in Agricultural Engineering), Universidade Federal do Vale do Sao Francisco, UNIVASF, Juazeiro-BA.

SALVIANO, A. A. C. **Variability of soil attributes and crotalaria juncea in degraded soil**

in the municipality of Piracicaba-SP. 91f. Thesis (Doctorate) - Escola Superior de Agricultura "Luiz de Queiroz", Universidade de Sao Paulo. Piracicaba, 1996.

SOHIDRA. Groundwater Directorate - DASUB. Available at: *http://www.sohidra.ce.gov.br/index.php/aguas-subterraneas.* Accessed on: October 10, 2016.

VASCONCELOS, S. M. S; SOUSA, A. J. G. de. **The Geostatistical Approach in Hydrogeological Studies: an Example of Application.** Revista de Geologia, Vol. 22, n° 1,61-74, 2009. Available at: www.revistadegeologia.ufc.br.

ZIMBACK, C. R. L. (2001). Spatial analysis of soil chemical attributes for soil fertility mapping purposes. **Thesis (Livre-Docência em Levantamento do solo e Fotopedologia)**. Faculty of Agronomic Sciences, São Paulo State University, Botucatu, 114f.

10. ANNEXES

Annex 1 - Parameters used for the depth indicator kriging map.

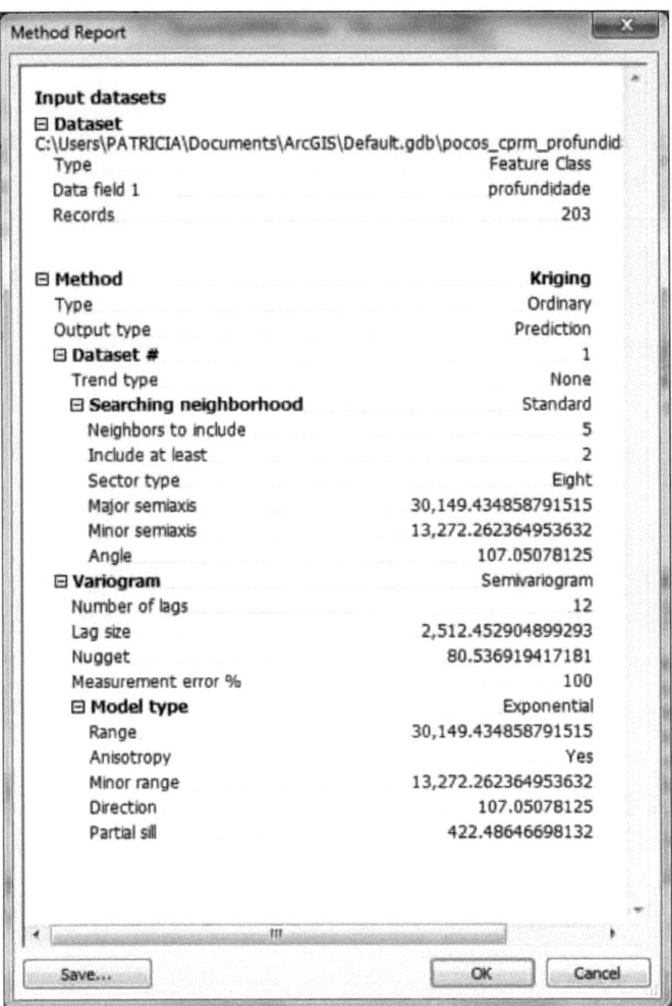

Method Report

Input datasets

Dataset
C:\Users\PATRICIA\Documents\ArcGIS\Default.gdb\pocos_cprm_profundid

Type	Feature Class
Data field 1	profundidade
Records	203
Method	**Kriging**
Type	Ordinary
Output type	Prediction
Dataset #	1
Trend type	None
Searching neighborhood	Standard
Neighbors to include	5
Include at least	2
Sector type	Eight
Major semiaxis	30,149.434858791515
Minor semiaxis	13,272.262364953632
Angle	107.05078125
Variogram	Semivariogram
Number of lags	12
Lag size	2,512.452904899293
Nugget	80.536919417181
Measurement error %	100
Model type	Exponential
Range	30,149.434858791515
Anisotropy	Yes
Minor range	13,272.262364953632
Direction	107.05078125
Partial sill	422.48646698132

Save... OK Cancel

Annex 2 - Parameters used for the static level indicator kriging map.

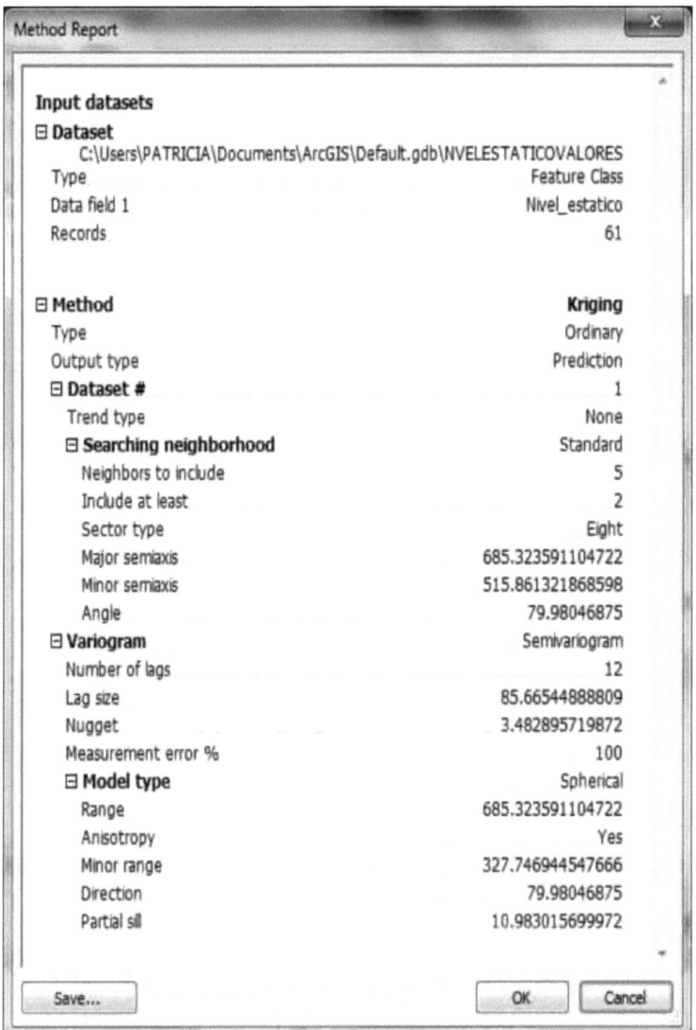

Method Report

Input datasets

Dataset	C:\Users\PATRICIA\Documents\ArcGIS\Default.gdb\NVELESTATICOVALORES
Type	Feature Class
Data field 1	Nivel_estatico
Records	61
Method	**Kriging**
Type	Ordinary
Output type	Prediction
Dataset #	1
Trend type	None
Searching neighborhood	Standard
Neighbors to include	5
Include at least	2
Sector type	Eight
Major semiaxis	685.323591104722
Minor semiaxis	515.861321868598
Angle	79.98046875
Variogram	Semivariogram
Number of lags	12
Lag size	85.66544888809
Nugget	3.482895719872
Measurement error %	100
Model type	Spherical
Range	685.323591104722
Anisotropy	Yes
Minor range	327.746944547666
Direction	79.98046875
Partial sill	10.983015699972

Save... OK Cancel

Annex 3 - Parameters used for the flow indicator kriging map.

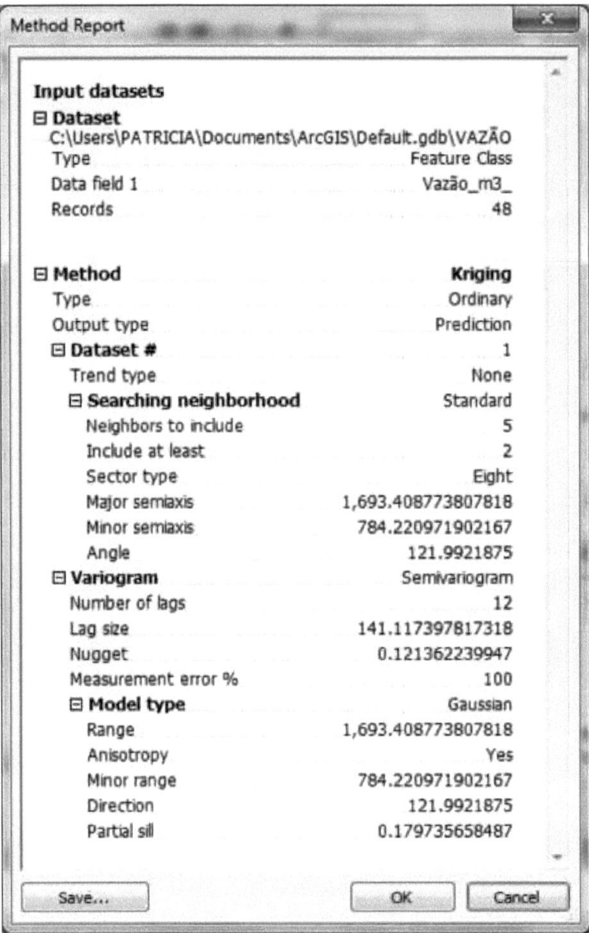

Method Report

Input datasets

⊟ **Dataset**	C:\Users\PATRICIA\Documents\ArcGIS\Default.gdb\VAZÃO
Type	Feature Class
Data field 1	Vazão_m3_
Records	48
⊟ **Method**	**Kriging**
Type	Ordinary
Output type	Prediction
⊟ **Dataset #**	1
Trend type	None
⊟ **Searching neighborhood**	Standard
Neighbors to include	5
Include at least	2
Sector type	Eight
Major semiaxis	1,693.408773807818
Minor semiaxis	784.220971902167
Angle	121.9921875
⊟ **Variogram**	Semivariogram
Number of lags	12
Lag size	141.117397817318
Nugget	0.121362239947
Measurement error %	100
⊟ **Model type**	Gaussian
Range	1,693.408773807818
Anisotropy	Yes
Minor range	784.220971902167
Direction	121.9921875
Partial sill	0.179735658487

Save... OK Cancel

I **want** morebooks!

Buy your books fast and straightforward online - at one of world's fastest growing online book stores! Environmentally sound due to Print-on-Demand technologies.

Buy your books online at
www.morebooks.shop

Kaufen Sie Ihre Bücher schnell und unkompliziert online – auf einer der am schnellsten wachsenden Buchhandelsplattformen weltweit! Dank Print-On-Demand umwelt- und ressourcenschonend produzi ert.

Bücher schneller online kaufen
www.morebooks.shop

info@omniscriptum.com
www.omniscriptum.com

Printed by Books on Demand GmbH, Norderstedt / Germany